AF715349

The Truth Of Reality

Also by The Quantum Shifter

The Truth of Reality
The Biggest Misconceptions About Manifestation
Why Isn't It Here Yet?
Manifest In Steps
Identity Shifting
One Decision Creates a New Reality
Self-Concept Selects Your Reality
Manifest Faster
Identity Shifting
I AM PERFECT
Embody a Wealth Identity
Mental Diet Mastery
Stop Spiraling While Manifesting
Emotional Mastery
Tell Yourself This When You're Depressed
Manifesting Money 101
Money From Nothing
You Are The Bank
Money Mindset For Single Moms
Your SP Loves You
Manifest A Celebrity
Have The Best Sex of Your Life
Lose Weight The Mental Way
Manifest Away "ADHD"
The Writer Everyone Reads
You Don't Need 3D Proof
Action Is Optional
The Positive Delusion

Cause and Effect Is a Lie
Objective Reality Is a Myth
Revision
Collapse & Command

The Truth of Reality

How to Claim the Only Law That Matters — Reality Is What You Say It Is

The Quantum Shifter

Identity Press

Connect with The Quantum Shifter

Website: www.TheQuantumShifter.com

Facebook: www.Facebook.com/TheQuantumShifter

ISBN: 978-1-972432-08-2

Published by Identity Press

These works were written over many years during the author's ongoing study of identity and reality creation.

They are now being published for the first time under the pen name *The Quantum Shifter*.

YOUR REALITY IS WHAT YOU SAY IT IS.

Contents

Before You Continue

This book will make the most sense if you are open to a simple idea: Your reality is not fixed. It is not happening to you. It is reflecting what you assume is true.

If you believe that reality is solid, unchangeable, and determined by circumstances, parts of this book may feel confusing or difficult to accept. And that's okay. You don't need to force yourself to believe anything. You can simply set this book down and return to it later, if and when it feels right.

But if something in you already senses that reality may not be what it seems—if you've noticed patterns that don't quite make logical sense—or if you've experienced moments where things shifted in ways you couldn't explain— then this book will likely feel familiar.

Not because it's teaching you something new, but because it's putting words to something you already know. Read it slowly. Let it land. And most importantly: Don't try to "do" anything. Just see what you notice.

“THE ONLY TRUTH THAT MATTERS
IS THIS:
YOUR REALITY IS WHAT YOU SAY
IT IS.”

Introduction: The One Truth Behind All Truths

People argue endlessly about how reality works. Some swear by the Law of Attraction. Others fight for the Law of Assumption. Some claim it's all karma, or destiny, or science, or God's will.

Scroll through books, videos, or forums and you'll see one thing in common: endless debates about which law is "right."

Here's the truth no one wants to admit:

They're all right — because reality molds itself to what they believe.

That's why two people can live by completely opposite rules and both find "proof" they're right. One person works hard and insists nothing comes without effort — and reality reflects that truth. Another person believes everything is effortless and falls into place — and reality reflects that truth too.

So which is correct? Both. And also, neither.

Because the only truth that matters is this:

Your reality is what you say it is.

Why This Book Exists

I didn't write this book to convince you of another system. You don't need more rules, steps, or rituals. You don't need to compare, debate, or argue about which "law" is supreme.

This book exists to strip everything back to the core truth: **your word is the only law.**

- When you say, "*This works for me*," it does.
- When you say, "*That doesn't apply to me*," it doesn't.
- When you say, "*My world operates on my terms*," it does.

It's that simple.

The Simplicity Most People Avoid

Why do people resist this truth? Because simplicity leaves no one to blame. If reality is what you say it is, there's no excuse, no waiting, no "if only."

It puts you in full authority. And authority feels uncomfortable – until you learn how to own it.

What You'll Learn

In this book, you'll discover how to:

- End the illusion of "objective" reality.
- See why all laws, systems, and rules bend to you.
- Treat beliefs like personal physics – and upgrade them at will.
- Choose from the buffet of reality without hesitation.
- Build core truths that stabilize your identity so reality has no choice but to reflect them.
- Apply this in daily life: relationships, money, success, and even the chaotic world around you.

Your Turning Point

This is not about hope. This is not about waiting for proof. This is not about figuring out which spiritual law is "right."

This is about living from the only truth that matters: **Reality is what you say it is.**

When you know this, life becomes simple. You say yes or no — and reality obeys.

This book will show you how to live from that freedom.

"If reality were objective, outcomes would be uniform. Instead, they bend to belief."

Chapter 1
The End of Objective Reality

For centuries, people have been chasing one question: *What is the true nature of reality?*

Philosophers, scientists, spiritual teachers, and everyday seekers have all tried to answer it. But the debates never end. Some insist reality is material and objective. Others claim it's spiritual and subjective. Still others argue it's a mix of both.

Here's the problem with all of that: they're trying to prove something that doesn't exist.

There is no objective reality.

Why Objective Reality Is a Myth

If reality were fixed, everyone would experience it the same way. But that's not what happens.

- Two people can live in the same city, under the same conditions — one experiences abundance, the other experiences struggle.

- Two people can try the same method — one swears it works every time, the other swears it doesn't.
- Two people can live through the same event — one calls it a blessing, the other calls it a curse.

If reality were objective, outcomes would be uniform. Instead, they bend to belief.

How People Prove Themselves Right

Look around and you'll notice: everyone can find evidence for their version of the truth.

- The person who believes "*hard work pays off*" finds success only through effort.
- The person who believes "*everything comes easily*" experiences flow and ease.
- The person who believes "*people can't be trusted*" continually attracts betrayal.
- The person who believes "*everyone supports me*" finds loyalty and help every-

where.

Opposite beliefs. Opposite results. And both are correct.

Not because reality is objective — but because reality is obedient.

Reality Is What You Say It Is

The foundation of this book, and the truth most people avoid, is this: **Your reality is what you think it is.**

There is no higher law, no absolute system, no universal rulebook. There is only your word — and the reality that reflects it.

This is why two people can live by completely opposite rules and both succeed. Reality doesn't care about the rules themselves. It only cares about who you've decided you are and what you've decided is true.

The Freedom in Knowing

At first, this truth can feel destabilizing. If there's no objective reality, what do you lean on? Where do you find certainty?

Here's the paradox: once you stop searching for "the one true law," you find the ultimate certainty — **the authority of your own word.**

- You no longer need to debate whether Law of Attraction or Law of Assumption is "more true."
- You no longer need to wait for science, scripture, or someone else's system to validate your path.
- You no longer need proof from the 3D world before declaring what's yours.

The only certainty you need is this: "*What I say goes.*"

Example: The Two Entrepreneurs

Two people start businesses at the same time.

- One believes success requires hustle and 80-hour workweeks. They live in constant stress — but eventually succeed, because their reality reflects their belief: "*Hard work pays off.*"
- The other believes success is natural, clients chase them, and money flows without effort. They build their business

in ease — and succeed too, because their reality reflects their belief: "*Everything is effortless.*"

Opposite truths. Identical success. Because both realities bent to the beliefs behind them.

Identity Statements

- "Objective reality doesn't exist. Reality obeys me."
- "Everyone is right — in their world. And I'm always right in mine."
- "What I say is the only foundation."
- "My word is the law of my reality."

Exercise: Breaking the Illusion of Objectivity

1. Write down three beliefs you've carried that felt "universally true." Examples: "*Love is hard.*" "*Money doesn't grow on trees.*" "*Success takes sacrifice.*"
2. Find someone in the world who clearly disproves each belief.
   - People who find love easily.

  - People who live in effortless wealth.
  - People who succeed without sacrifice.

3. Recognize: these beliefs were never universal laws. They were just stories you accepted.
4. Declare your new truth: "*Reality obeys what I say, not what others believe.*"

Closing Reminder

Objective reality is the biggest illusion. The truth is simpler, sharper, and more freeing: **your reality is what you say it is.**

This is the only foundation you'll ever need — and once you claim it, you'll never waste another second debating "which law is right."

Because you'll know: *the law is you.*

Integration Notes

“Different laws.
Same outcome: they’re right
because reality mirrors their
chosen belief.”

Chapter 2

Laws That Bend to You

If you've ever searched for answers about manifestation, you've probably come across endless "laws":

- The Law of Attraction.
- The Law of Assumption.
- The Law of Vibration.
- The Law of Cause and Effect.
- Karma.
- Destiny.
- Fate.
- Even religious rules that claim to govern life.

Every system insists *their law is the ultimate truth*. But look closely, and you'll notice something strange: **all of them work.**

Why All Laws Work (and Why None of Them Are Ultimate)

- People who swear by the **Law of Attraction** do see results when they focus on "positive energy."
- People who use the **Law of Assumption** create exactly what they assume.
- People who live by **karma** constantly experience life as if actions circle back.
- People who believe in **destiny** often walk into opportunities that "prove" it's all pre-written.

How can all of these contradictory systems work?

Because they don't work on their own. **They work because the person believes in them.**

Reality Bends to Your Chosen Law

Think of "laws" like apps on your phone. You can install one, run it, and it will function exactly as described. But the phone itself is not the app.

Reality is the phone. The "laws" are just apps you run because you chose them.

- If you say, "*The Law of Attraction works for me*," it does.
- If you say, "*The Law of Assumption is the only way*," it is.
- If you say, "*Only karma matters*," then karma rules your reality.

It's not the law that has power. It's you.

Real-World Examples

- **The Effortless Creator:** One woman swears by Law of Attraction, makes vision boards, and feels good daily. Her life reflects it — opportunities, love, and wealth show up because she expects them.
- **The Assumption Purist:** Another person mocks vision boards but lives fully by Law of Assumption. They affirm, "*I'm always chosen, money flows easily to me*" — and it works, because they assume it does.
- **The Karma Believer:** A man insists karma governs everything. He gives generously, and life reflects generosity back. He also fears punishment, and sure

enough, he experiences setbacks – not because karma is "objective," but because he lives as if it is.

Different laws. Same outcome: they're right because reality mirrors their chosen belief.

Why Debates Don't Matter

When people argue about which law is true, they miss the point. It's like two people fighting over whether Apple Maps or Google Maps is "the real map." Both work. Both will get you where you want to go.

The power isn't in the app. It's in the phone.

The power isn't in the law. It's in you.

Identity Statements: Authority Over Laws

- "No law rules me. My word is the law."
- "Every system works because I say it does."
- "I choose what governs my reality."
- "Reality bends to me, not the other way around."

Exercise: Declaring Your Law

1. Write down all the "laws" you've believed in: attraction, assumption, karma, destiny, etc.

2. For each one, ask: *Did it work because it was ultimate truth, or because I decided it was true?*

3. Circle the moments you proved yourself right — and realize: you were the power all along.

4. Write this declaration: "*My reality is not governed by laws outside me. Reality is what I say it is.*"

Closing Reminder

Every "law" is optional. Every system works for someone. And every debate about which is correct misses the point entirely.

The only law that matters is the one you claim: **Reality is what you say it is.**

Integration Notes

"THE MOMENT YOU CHANGE THE BELIEF, YOU CHANGE THE LAW THAT GOVERNS YOUR REALITY."

Chapter 3

Beliefs Become Physics

If you've ever wondered why your life seems to run on certain "rules," the answer is simple: **your beliefs have hardened into laws.**

Beliefs act like personal gravity. They make your world behave in ways that seem consistent and inescapable. To you, they feel like unshakable truths. But they are not universal truths. They are chosen truths that have become your private physics.

How Beliefs Turn Into Physics

- If you believe "*money is hard*," then in your world, money really does show up as hard. Bills pile up, income feels slow, and effort seems required.

- If you believe "*love never lasts*," then relationships confirm that story until it feels like proof.

- If you believe "*I'm always lucky*," then life arranges itself to keep delivering evidence of your luck.

In each case, ***the belief has become your law.*** *It functions exactly like physics: it governs how reality behaves around you.*

Why Beliefs Feel So "True"

The reason beliefs feel so convincing is because reality reflects them back with consistency.

- Believe in scarcity, and you'll see shortage everywhere.
- Believe in abundance, and you'll see opportunities multiply.
- Believe you are overlooked, and people ignore you.
- Believe you are magnetic, and people notice you instantly.

It's not that one belief is more real than another. It's that reality is obedient. It delivers evidence that matches your declared identity.

Upgrading Your Physics

The power in knowing this is simple: *if your beliefs have become your personal physics, then you can rewrite those beliefs and create new laws.*

- Scarcity → "*I've always had more than enough.*"
- Rejection → "*I've always been chosen.*"
- Struggle → "*Things always work out easily for me.*"
- Chaos → "*My world is calm, orderly, and supportive.*"

When you stabilize these upgraded beliefs, reality stabilizes with them. The new belief becomes the new physics.

Everyday Examples

Money Example One person grows up hearing, "*You have to struggle to succeed.*" Their life becomes proof of that law: long hours, slow growth, constant obstacles. Another person grows up hearing, "*Money flows easily in our family.*" Their life reflects effortless wealth. Neither law is ultimate. Each one is simply physics created by belief.

Relationships Example A woman believes, "*Men always leave.*" Every relationship confirms it, no matter the circumstances. Another woman believes, "*I am always loved and chosen.*" Her partners reflect devotion and consistency. The difference isn't circumstance—it's physics.

Identity Statements

- "My beliefs are my physics."
- "When I change my beliefs, I change my reality."
- "My world always reflects the identity I stabilize."
- "Nothing is fixed; my story sets the law."

Exercise: Identifying and Rewriting Your Physics

1. Write down three "laws" you believe about yourself or your life. Examples: "*I always have to work hard,*" "*I'm not good with money,*" "*People take advantage of me.*"
2. Ask yourself: *Has reality consistently reflected this belief back to me?*

3. Write the upgraded version of each law: *"Things are always easy for me," "I've always been wealthy," "People always treat me with respect."*

4. Declare these daily until they feel like your natural identity.

Closing Reminder

Your beliefs are not harmless thoughts. They are the physics of your world. They create consistency, not because they are objectively true, but because reality has no choice but to reflect them.

And the moment you change the belief, you change the law that governs your reality.

Integration Notes

"NOTHING BECOMES YOUR REALITY UNTIL YOU SAY YES AND PUT IT ON YOUR PLATE."

Chapter 4

The Buffet Metaphor

Reality can feel overwhelming because it seems like everything is coming at you at once: news headlines, other people's opinions, past experiences, and potential futures. But the truth is much simpler.

Life is like a buffet.

At a buffet, you don't eat everything. You only eat what you put on your plate. The food that isn't chosen has no effect on you. In the same way, reality constantly offers you options, but you only experience the ones you claim.

Everything Exists at Once

At the buffet of life, every possible story is available:

- Stories of wealth, success, and abundance.
- Stories of struggle, poverty, and lack.

- Stories of love, stability, and devotion.
- Stories of heartbreak, betrayal, and loss.

All of it is there, side by side. But nothing becomes *your* reality until you say yes and put it on your plate.

Choosing Consciously

Most people wander through life piling things on their plate without noticing:

- They pick up scarcity stories from their family.
- They add fear stories from the news.
- They spoon on doubt from past failures.

Then they wonder why their plate looks unappetizing.

Revision begins with recognizing the buffet for what it is: a spread of options. Your job is not to argue about what's on the table. Your job is to choose what you want to eat.

The Power of Saying No

At a buffet, you don't waste time protesting the dishes you don't like. You don't yell at the chef for putting out food that isn't to your taste. You simply say no and walk past it.

Reality works the same way.

- See a headline you don't want? Say, "*Not in my reality.*"
- Hear someone predict struggle? Say, "*That doesn't apply to me.*"
- Notice an old story rising? Say, "*No, that's not my plate.*"

You don't fight it. You don't debate it. You simply decline it.

Everyday Examples

Relationships You hear someone say, "*All relationships fall apart eventually.*" At the buffet of reality, that dish is available — but you don't have to put it on your plate. Instead, you choose: "*Love in my world is lasting, stable, and fulfilling.*"

Money Your parents lived by, "*Money is hard to come by.*" That story is on the table, but you don't have to serve it. You choose: "*Money is easy. Wealth is natural for me.*"

Health Society says, "*Aging means decline.*" That's a dish being served. You can decline it and choose: "*I grow stronger, healthier, and more vibrant each year.*"

Identity Statements

- "Life is a buffet, and I choose what goes on my plate."
- "I don't argue with the food I don't want. I simply say no."
- "What I put on my plate becomes my lived reality."
- "Everything exists, but only my choices matter."

Exercise: Your Plate of Reality

1. Write down three stories you've "put on your plate" without realizing it. Example: "*Money is scarce,*" "*Love is hard,*" "*I always get overlooked.*"

2. Cross them out and write: "Not *on my plate.*"
3. Replace each one with the story you now choose: "*Money is abundant,*" "*Love is stable,*" "*I am always valued.*"
4. Repeat until choosing consciously becomes second nature.

Closing Reminder

The buffet of reality is endless. Everything exists at once. But only what you put on your plate becomes your truth.

Stop arguing about the dishes you don't want. Stop piling your plate with stories that don't serve you.

Choose deliberately. Because in your world, only your choices matter.

Integration Notes

"REALITY OBEYS YOUR AUTHORITY IN REAL TIME."

Chapter 5

Saying Yes or No Instantly

Most people hesitate when faced with a choice of reality. They see a story, a situation, or a headline and freeze: "*Is this true? Do I need to worry? Should I fight it? What if it affects me?*"

That hesitation is doubt. And doubt feeds the very story you don't want.

The truth is simpler: reality responds to your immediate decision. Yes or no. Accept or decline. Claim or dismiss.

Why Instant Decisions Matter

Reality doesn't wait for debate. It reflects your word. The longer you linger in questioning, the more you accidentally stabilize the unwanted story.

- The moment you say yes, reality collapses into that truth.

- The moment you say no, it dissolves.

There is no negotiation. There is no delay. Reality obeys your authority in real time.

The Practice of Yes and No

1. **See the Story.** A comment, a news headline, a thought from the past.
2. **Respond Instantly.**
    - If it matches your chosen reality: "*Yes, that's mine.*"
    - If it doesn't: "No, *that's not in my reality.*"
3. **Move On.** No need to argue, prove, or analyze. Your yes or no is final.

Everyday Examples

Relationships A friend says, "*All men cheat.*" Old you might debate or feel fear. New you simply says: "Not *in my reality.*" Done.

Money You see a headline: "*Economic collapse coming.*" Old you would spiral. New you says: "Not *in my world. In my world, money flows easily.*"

Health Someone comments, "*Everyone gains weight with age.*" Old you absorbs it. New you says: "Not *true for me. I grow healthier every year.*"

Why This Is So Powerful

Most people believe they need to explain, argue, or justify their stance. They don't. The power isn't in explaining. It's in declaring.

When you say yes or no instantly, you cut off the spiral before it begins. **You collapse reality into the story you've claimed, with no delay and no leakage of power.**

Identity Statements

- "My yes is law. My no is law."
- "I decide instantly. Reality obeys instantly."
- "I don't debate. I declare."
- "The moment I choose, the story is mine or gone."

Exercise: Instant Response Training

1. For the next 24 hours, notice every story, headline, or comment that comes across your plate.

2. Respond instantly: "Yes, that's mine" or "No, not in my reality."

3. Don't linger, argue, or explain. Practice moving on the moment you decide.

4. At the end of the day, write down how much calmer and clearer you feel.

Closing Reminder

You are not at the mercy of every story life serves you. You are the chooser. And the fastest way to stabilize your authority is this: yes or no, instantly.

What you say yes to becomes yours. What you say no to disappears.

Nothing more is required.

Integration Notes

"Reality isn't shaped by what's "objectively true." It's shaped by what you declare as true."

Chapter 6

The Clickbait Illusion

We live in a world of clickbait. Headlines scream for attention, stories are exaggerated, and people argue about what's true or false. The reflex most people develop is skepticism: "*Let me double-check. Let me fact-check. Let me see if this source is reliable.*"

It seems like a smart habit. But here's the hidden problem: skepticism is just another way of saying, "*I don't know if I have authority here.*"

And the longer you hesitate, the more you feed doubt into your reality.

How Skepticism Weakens You

- When you "check the source," you're putting the authority outside yourself.
- When you wait for proof, you're delaying your word.
- When you debate whether something is

true, you're reinforcing the possibility that it could be.

Skepticism doesn't protect you. It stabilizes uncertainty.

Why Sovereignty Is Stronger

Reality isn't shaped by what's "objectively true." It's shaped by what you declare as true.

- The source doesn't matter.
- The headline doesn't matter.
- The statistics don't matter.

What matters is whether you say *yes* or *no*.

Everyday Examples

Money You see a financial article: "*Recession is inevitable.*" The skeptic digs for data. The sovereign simply says: "Not *in my world. Money always flows to me.*"

Health You read a report: "*Most people over forty face decline.*" The skeptic checks studies. The sovereign declares: "Not *true for me. I grow healthier every year.*"

Relationships You hear gossip: "*Everyone cheats eventually.*" The skeptic asks around. The sovereign decides: "Not *in my reality. My relationships are loyal and secure.*"

Why Doubt Is the Real Clickbait

Clickbait only works because it hooks attention. Doubt works the same way: it lures you into questioning, feeding energy into a story you never wanted to claim.

The more you research, the more you argue, the more you "wait to see" — the more that story grows roots in your world.

But the moment you say "No, *that's not mine,*" the clickbait dissolves.

Identity Statements

- "I am the source. Nothing outside me decides what's true."
- "Skepticism is delay; sovereignty is authority."
- "I don't need proof. My word is proof."
- "I declare reality, I don't debate it."

Exercise: Declaring Over Clickbait

1. Scroll through news or social media for five minutes.
2. Each time a headline tries to grab your attention, stop and say:
   - "Yes, that's mine," or
   - "No, not in my reality."
3. Notice how quickly your energy feels clean compared to the heaviness of skepticism.
4. Repeat daily until instant sovereignty becomes natural.

Closing Reminder

Clickbait thrives on keeping you hooked in doubt. But you are not a skeptic – you are sovereign.

Your authority doesn't depend on the source, the headline, or the data. Your authority is your word.

And once you claim that, reality has no choice but to obey.

Integration Notes

"HERE'S THE TRUTH:
THE PEOPLE IN YOUR WORLD ARE
MIRRORS OF YOUR IDENTITY."

Chapter 7

Identity = Projection

Most people believe they are interacting with separate, independent individuals. They assume others have their own agendas, thoughts, and choices that operate outside of them.

But here's the truth: the people in your world are mirrors of your identity.

They cannot show up in ways that contradict who you've declared yourself to be.

Why People Reflect You

Reality is not random. Every encounter, every conversation, every reaction is filtered through the identity you hold.

- If you believe "*I'm overlooked*," people forget you, ignore you, or dismiss your words.
- If you believe "*I'm magnetic*," people no-

tice you, listen to you, and want more of your presence.

- If you believe "*I'm always chosen*," your friends, partners, and clients confirm that truth without fail.

It looks like they're acting on their own. But in your world, they can only act according to your stabilized identity.

How This Shows Up in Everyday Life

Specific Person (SP) If you see your SP as distant or unavailable, they mirror that story back. Revise them in your mind as loving, devoted, and present – and they must reflect it.

Friends and Family If you believe you're the "responsible one" who always carries the weight, friends and family lean on you. If you shift to "*I am supported and taken care of*," those same people begin showing up differently.

Clients and Business If you believe clients are hard to find, they avoid you. If you believe "*clients chase me down because I'm the best at what I do*," suddenly people appear, eager to work with you.

Strangers If you believe "*people are kind to me,*" strangers hold doors, smile, and offer help. If you believe "*the world is hostile,*" strangers seem rude or impatient.

Stories from Real Shifts

- A woman who always felt "ignored" in groups revised herself as magnetic. Within weeks, she noticed people turning toward her, asking her opinion, and inviting her into conversations without her saying a word.
- A man who struggled with clients revised his identity to "*I am the obvious choice.*" Suddenly, prospects who had gone silent resurfaced, saying they felt "drawn back" to him.
- A woman who carried the wound "*men leave me*" revised her identity to "*I've always been loved and chosen.*" Her relationships stabilized, and she now describes her love life as "effortlessly secure."

Identity Statements

- "Everyone in my world is a reflection of me."
- "I decide who I am, and others must reflect it."
- "I am magnetic, chosen, and valued – and my world shows me that truth."
- "There is no separation; reality mirrors my identity."

Exercise: Rewriting a Reflection

1. Think of one person who isn't showing up the way you'd like.
2. Ask yourself: "*What story am I holding about them? What identity of mine are they reflecting?*"
3. Revise the story. Write down the new identity: "*I am always respected. I am always chosen. People reflect devotion to me.*"
4. Imagine a simple scene where that person treats you exactly as your new identity says.

5. Repeat until it feels natural. Then watch how they begin to shift.

Closing Reminder

The people in your world are not exceptions. They are mirrors. They cannot resist your stabilized identity.

When you change the story you hold about yourself, you change the way everyone shows up.

Because in your world, identity is projection.

Integration Notes

"Reality is not collective. Everyone lives in their own reflection."

Chapter 8

Acceptance Without Agreement

One of the biggest traps in reality creation is the urge to argue.

You've probably noticed how easily people get pulled into debates:

- Religion vs. atheism.
- Science vs. spirituality.
- Law of Attraction vs. Law of Assumption.
- Positive thinking vs. realism.
- Politics, culture, world events.

Everyone wants their version of reality to be "the truth." And if you're not careful, you'll waste your energy trying to convince others of what you know.

But here's the freedom: **you don't have to agree, and you don't have to argue.**

Why You Don't Need Agreement

Reality is not collective. Everyone lives in their own reflection.

- If someone says, "*God controls everything,*" that's true in their world.
- If someone says, "*Science is all that matters,*" that's true in their world.
- If someone says, "*I create my reality with my assumptions,*" that's true in their world.

And if you say, "*Reality is what I say it is,*" that's true in yours.

There is no conflict unless you create it.

The Power of Letting Others Be Right

Most people waste their energy fighting to be "more correct." But what if you let everyone be right in their own world?

This doesn't weaken you. It strengthens you. Because while they argue, you stand steady in the authority of your own law.

Everyday Examples

Family and Religion Your parents may say, "*Everything is in God's hands.*" Old you would argue. New you accepts: that's their buffet plate, not yours. Your truth still stands.

Friends and Politics Your friend rants, "*The world is falling apart.*" Old you would panic or debate. New you accepts: that's their reality. In yours, the world supports you.

Co-workers and Struggle A co-worker says, "*You have to work yourself to death to succeed.*" Old you might defend yourself. New you accepts: that's their truth. In yours, success comes easily.

How This Creates Peaceful Power

When you stop fighting for agreement, you step into calm, secure authority.

- You no longer react to fear-based conversations.
- You no longer feel the need to prove your system works.
- You no longer spiral when someone insists on their version of reality.

Instead, you embody the truth: "*They're right for them. I'm right for me.*"

Identity Statements

- "I don't need agreement to be right."
- "Others are right in their world. I'm right in mine."
- "I allow others their buffet while I enjoy mine."
- "My truth doesn't require debate. My word is final."

Exercise: The Next Debate

1. Think of someone in your life who often argues their beliefs (family, friend, co-worker).
2. Next time they share their version of reality, pause.
3. Instead of debating, silently affirm: "*That's their world, not mine.*"
4. Walk away without engaging, and feel the peace of not needing to win.

Closing Reminder

Authority doesn't need agreement. Sovereignty doesn't need defense.

The more you accept that others are right in their world, the more secure you become in your own.

And in that security, your reality reflects nothing but your chosen truth.

Integration Notes

"Core truths are simple declarations that define how your reality operates."

Chapter 9

Building Your Own Core Truths

At this point, you've seen the principle: reality isn't objective, and no outside law governs you. The only law that matters is what you say.

But here's the challenge: if you don't consciously choose your truths, you'll unconsciously repeat the ones you inherited — from family, culture, or past experiences.

That's why building your own core truths is essential. They become the anchors that stabilize your identity and keep your world reflecting exactly what you've chosen.

What Are Core Truths?

Core truths are simple declarations that define how your reality operates. They are not affirmations you repeat hoping they'll come true. They are decisions you claim as final.

Examples:

- "What I say goes."
- "Everything in my world supports me."
- "I am perfect, and my world reflects perfection."
- "Everyone's me pushed out."
- "My word is the only authority in my life."

These are not techniques. They are foundations.

Why Core Truths Matter

When you stabilize a few core truths, you no longer wobble when challenges appear.

- If a bill arrives, you fall back on: "I *am always provided for. Money flows easily to me.*"
- If someone argues with you, you rest in: "*They are right in their world, and I'm right in mine.*"
- If an old memory surfaces, you anchor into: "*The past has no authority. My word is final.*"

Without core truths, you risk reacting to every trigger. With core truths, you stay grounded no matter what.

How to Create Your Own Core Truths

1. **Identify Old Stories** Write down the limiting "laws" you've lived under. Example: "*Nothing comes easy*," "*People always leave me*," "*Money is scarce*."
2. **Flip Them Into Permanent Truths** Cross them out and replace them with final declarations. Example:
    - Old: "*Nothing comes easy*." → New: "*Everything works out easily for me*."
    - Old: "*People always leave me*." → New: "*I've always been loved and chosen*."
3. **Choose 3–5 Anchors** Pick the truths that feel strongest. Keep them simple, powerful, and final.
4. **Stabilize Them Daily** Repeat them, write them, embody them until they feel like your permanent physics.

Everyday Examples

Relationships Old story: "*Love is unpredictable.*" Core truth: "*Love in my world is stable, loyal, and lasting.*"

Money Old story: "*Money slips through my fingers.*" Core truth: "*Money sticks to me and multiplies easily.*"

Success Old story: "*I always fall short.*" Core truth: "*I always win. My success is inevitable.*"

Identity Statements

- "My reality is anchored by my chosen truths."
- "I decide once, and reality obeys."
- "I need only a few strong truths to govern my world."
- "What I declare as law becomes my permanent physics."

Exercise: Writing Your Core Truths

1. Write down 5 limiting "laws" you've believed.

2. Cross them out with a bold line.

3. Rewrite each one as a permanent declaration.

4. Choose 3–5 to stabilize as your personal core truths.

5. Say them daily until they feel natural, obvious, and unshakable.

Closing Reminder

Core truths simplify life. You don't need hundreds of affirmations, rituals, or techniques. **You need a handful of permanent decisions about how your world works.**

And once you stabilize those truths, reality has no choice but to reflect them — consistently, automatically, and without effort.

Integration Notes

"When you claim truths of
being loved, chosen,
and respected,
your relationships have no
choice but to reflect them."

Chapter 10
Relationships

Relationships are one of the clearest mirrors of your chosen reality. **Every argument, every moment of closeness, every pattern that repeats is nothing more than a reflection of the truths you've stabilized.**

When people struggle in relationships, it's not because love is inherently difficult. It's because they've unconsciously chosen truths that make it so.

Why Relationships Feel Like Battles

Most relationships fall into conflict because two people are pulling from different buffets:

- One person says, "*Love is hard work.*"
- The other says, "*Love should be effortless.*"
- They spend their time trying to convince each other their buffet is the "real" one.

Arguments are not about facts. They are battles over whose truth will dominate.

Choosing Partners Who Reflect You

When you stabilize your own truths, your partners and friends have no choice but to reflect them.

- If you stabilize: "*I am always loved and chosen*," your relationships mirror loyalty and devotion.
- If you stabilize: "*People treat me with respect*," you stop attracting partners who belittle you.
- If you stabilize: "*Love is easy and fulfilling*," love stops feeling like drama and starts feeling like peace.

The truth is simple: you don't get what you want, you get who you are.

Letting People Upgrade or Exit

What about people who don't match your chosen truths? They have two options:

1. Upgrade – rise into alignment with your stabilized identity.

2. Exit — fade out of your world completely.

This isn't personal. It's projection. People cannot resist your stabilized identity. They either adapt to it or disappear.

Everyday Examples

Romantic Relationships Old truth: "*I always get left behind.*" New truth: "*I've always been loved and chosen.*" Result: Partners stop drifting away, and love reflects stability.

Friendships Old truth: "*Friends use me and never give back.*" New truth: "*My friendships are balanced. I am supported as much as I give.*" Result: Old one-sided friendships fade, and new supportive friends appear.

Family Old truth: "*My family never respects me.*" New truth: "*I am respected and valued in every relationship.*" Result: Conversations shift, and family members begin treating you differently.

Identity Statements

- "I am always loved, chosen, and respected."
- "My relationships reflect loyalty, peace, and devotion."
- "People in my world rise to match my truth or exit."
- "Love is easy because I say it is."

Exercise: Revising a Relationship Story

1. Write down one repeating pattern in relationships. Example: "*I always get ignored.*"
2. Cross it out and write your new core truth. Example: "*I've always been seen, heard, and valued.*"
3. Each time you notice the old pattern, declare: "*That's not my story. My reality is [new truth].*"
4. Watch as the people in your world adjust to reflect your stabilized identity.

Closing Reminder

Relationships are not mysterious. They are mirrors.

When you claim truths of being loved, chosen, and respected, your relationships have no choice but to reflect them.

And when you stop fighting over whose buffet is real, you live in the only truth that matters: your own.

Integration Notes

"Here's the truth: money is not objective. Success is not objective. Both bend to your chosen truths."

Chapter 11

Chapter 11: Money and Success

Few areas trigger more fear and debate than money. Some say you must grind and sacrifice. Others insist it's about mindset and flow. Some point to "the economy" or "luck."

Here's the truth: money is not objective. Success is not objective. Both bend to your chosen truths.

Scarcity vs. Abundance Are Just Stories

Scarcity feels real because you've stabilized it as a law. Abundance feels real because you've stabilized it as a law.

- Believe "*money is hard*," and your life confirms it.
- Believe "*money flows easily*," and your life confirms it.

- Believe "*success takes sacrifice*," and you'll sweat for every inch.

- Believe "*success comes naturally*," and opportunities fall into your lap.

Neither story is ultimate. Each becomes a private physics because you declared it so.

Declaring Your Money Truth

Instead of reacting to old stories – family patterns, economic fears, cultural scripts – decide what's true for you.

Examples:

- "*I've always been wealthy. Money flows naturally to me.*"

- "*Clients chase me down because I'm the best at what I do.*"

- "*My success is inevitable and effortless.*"

- "*Every dollar I spend multiplies back to me.*"

When you stabilize truths like these, money and success stop being a battle and start being automatic.

Everyday Examples

Career Old truth: "*Promotions are rare and competitive.*" New truth: "*I am the obvious choice. My career rises easily.*" Result: Raises, recognition, and opportunities flow without struggle.

Business Old truth: "*Clients are hard to find.*" New truth: "*Clients chase me down. My work is always valued.*" Result: Ideal clients appear with little effort.

Windfalls Old truth: "*Money only comes through hard work.*" New truth: "*Money surprises me constantly. Windfalls are normal.*" Result: Refunds, bonuses, and unexpected income show up.

Success Story

A man once told me, "*I always have to work twice as hard as everyone else just to get ahead.*" That story had become his private law. He decided to revise it to: "*I always succeed effortlessly. Everything I touch multiplies.*"

Within weeks, he noticed small shifts: less resistance at work, tasks finishing faster, colleagues offering help. Within months, he was promoted into a role that required less effort but paid more.

The work hadn't changed. His identity had. And reality reorganized to reflect it.

Identity Statements

- "I've always been wealthy."
- "Money flows easily and naturally to me."
- "My success is inevitable."
- "I am the best at what I do, and the world reflects it."
- "Every money story bends to the truth I declare."

Exercise: Choosing Your Wealth Truth

1. Write down the three strongest money beliefs you've carried. Example: "*Money is scarce,*" "*I have to struggle,*" "*I never get ahead.*"
2. Cross them out boldly.
3. Replace them with permanent declarations. Example: "*I've always been*

wealthy,” “Everything is easy for me,” “I always rise.”

4. Say them daily until they feel natural. Watch how money and success begin to mirror them.

Closing Reminder

Money and success are not objective forces. They are stories — and whichever one you stabilize becomes your lived reality.

When you choose abundance, success, and overflow as your permanent truths, your world has no choice but to obey.

Integration Notes

"The collective is not your authority.
Your word is."

Chapter 12

The World at Large

Turn on the news, scroll social media, or listen to conversations in public, and you'll hear endless declarations about how "the world is."

- *"The economy is collapsing."*
- *"The planet is in crisis."*
- *"Society is broken."*
- *"People are getting worse."*

It can feel overwhelming. But remember: these are dishes on the buffet. Just because they're loud doesn't mean you have to put them on your plate.

Collective Beliefs Are Optional

Most people assume that "collective reality" is unavoidable. They believe if the world agrees something is true, they must live under it.

But you've already seen that's not how reality works.

- If the world says "*money is scarce*," you can still live in abundance.
- If the world says "*relationships always fail*," you can still live in devotion.
- If the world says "*chaos is everywhere*," you can still live in peace.

The collective is not your authority. Your word is.

Choosing Your Parallel World

When chaos is loud, you have two options:

1. Argue and get pulled into the noise.
2. Decline it and stabilize your own world.

Each time you say, "*That's not in my reality*," you shift into the parallel world where your chosen truth is the only law.

Everyday Examples

Economy Headline: "*Recession is inevitable.*" Old self spirals. New self says: "Not *in my world. My wealth is stable and increasing.*"

Culture People insist, "*Dating is hopeless these days.*" Old self worries. New self declares: "*Love is easy and fulfilling in my world.*"

Health Trends Society says, "*Everyone declines with age.*" Old self accepts it. New self decides: "*I grow healthier, stronger, and more vibrant every year.*"

Why This Creates Calm in Chaos

The more chaotic the collective seems, the more authority you have when you don't waver.

Instead of fear, you feel peace. Instead of doubt, you feel certainty. Instead of reacting to the crowd, you stand in sovereignty: "*They are right in their world. I am right in mine.*"

Identity Statements

- "The collective does not rule me."
- "My parallel world reflects only my chosen truths."
- "I am calm and secure no matter what the world says."
- "Noise fades; my word remains."

Exercise: Rewriting the News

1. Watch or read five headlines.
2. For each, ask: "Do I *want this on my plate?*"
3. If no, say: "Not *in my reality.*"
4. Replace it with your declaration. Example:
    - Headline: "*Economic collapse.*"
    - Declaration: "My *wealth is stable and growing.*"

Do this for one week. Notice how your sense of calm and authority increases, even when chaos rises.

Closing Reminder

The world at large is not your master. Collective stories are not binding laws.

You are not here to agree with the noise. You are here to choose your plate.

And when you stand in your chosen truths, the chaos of the collective fades – leaving only the peace of the world you've declared.

Integration Notes

"What you say yes to becomes your reality.
What you say no to disappears."

Conclusion: The Freedom of Knowing

At every turn, you've seen the same principle repeated in different forms.

- There is no objective reality.
- No universal law governs you.
- No collective story can bind you.

Reality is what you say it is.

Why This Truth Frees You

Most people live in endless confusion. They study new laws, debate systems, and chase proof. They wait for someone else to tell them what's real.

But when you live from the truth of reality, all of that falls away. You stop chasing, doubting, and waiting. You stop arguing about who's right.

Because you know: everyone is right in their world, and you are right in yours.

The Simplicity of Yes and No

Life is no longer about analysis. It's about decision.

- See a story. Say yes or no.
- Feel a thought. Say yes or no.
- Hear a belief. Say yes or no.

What you say yes to becomes your reality. What you say no to disappears.

That's it. The complexity was never necessary.

Living From Core Truths

When you stabilize your own truths, life becomes predictable in the best way.

- Your relationships mirror loyalty and devotion.
- Your money reflects abundance and overflow.
- Your health reflects strength and vitality.

- Your world reflects peace, no matter what others claim.

Because your chosen truths become your physics, and reality cannot resist them.

Your Final Authority

This book isn't about giving you another system. It's about returning the authority to you.

The truth of reality is not a philosophy to argue over. It is a way of living. The simplicity is unshakable: **your word is the only law.**

Closing Declaration

Stop debating. Stop doubting. Stop waiting for proof.

Stand in the only truth that matters: **Reality is what you say it is.**

And once you know this, you are free.

About The Author

The Quantum Shifter is the pen name of an Ivy-trained psychologist, researcher, and lecturer whose work explores the intersection of identity, consciousness, and reality creation. After completing advanced studies in psychology and spending years teaching and writing at the university level, she shifted her focus toward the deeper question that shaped her life's work:

How does identity create reality — and why does changing the story of who you are change the world you live in?

Her publications draw on decades of private research, academic inquiry, and experiential observation. Known for her clarity, precision, and deeply grounded approach, she writes about identity physics, timeline selection, self-concept, and the mechanics of reality in a way that is both accessible and transformative.

Although her academic background is traditional, her pen name reflects the evolution of her work – a nod to the moment she realized that consciousness itself is the most powerful instrument of change. The name stuck, and readers around the world now associate *The Quantum Shifter* with high-level insight, elegant simplicity, and rapid identity transformation.

She currently lives a quiet, contemplative life dedicated to writing, research, and teaching the principles of identity-based reality creation. Her books serve as distilled explorations of the concepts she has spent decades studying – written for readers who are ready to shift who they are, not just what they want.

www.ingramcontent.com/pod-product-compliance
Lightning Source LLC
Chambersburg PA
CBHW022105240926
51848CB00008B/149
9781972432082